QUAIL FARMING UNLEASHED

Raise Quails for Eggs and Meat at Home

The Ultimate Guide To Quail Husbandry For Sustainable And Profitable Farming

Dr. Fabian Felicity

Table of Contents

CHAPTER ONE

Introduction

Quail farming, a lesser-known but more popular business in the poultry farming industry, has gained popularity due to its multiple benefits and minimal care needs.

As the need for alternative protein sources grows, quail farming becomes a realistic and lucrative choice for both small and large-scale farms. This page gives a full review of quail farming, including the fundamentals, choosing the best quail breed, and constructing a perfect environment for these little birds.

The Basics Of Quail Farming

Quails, members of the Phasianidae family, are tiny game birds noted for their delicious and nutritious meat and eggs.

Unlike conventional poultry such as chickens and ducks, quail farming has distinct advantages, making it an appealing alternative for farmers seeking variety in their agricultural interests. The most popular species produced for commercial reasons are Coturnix quail, Bobwhite quail, and Pharaoh quail.

One of the key benefits of quail farming is that these birds mature rather quickly. Quails mature in around six weeks, giving them a reliable supply of meat and eggs. Furthermore, quail are noted for their tolerance to many temperatures and habitats, making them appropriate for agricultural operations in a variety of locations.

Quail eggs, which are smaller in size and have a unique taste, have become popular among health-conscious customers. Quail eggs are a healthier alternative to chicken eggs due to their high protein, vitamin, and mineral content. Furthermore, quail meat is thin and tasty, appealing to people seeking a new gastronomic experience.

Choosing The Right Quail Breed For Your Farm

Choosing the right quail breed is an important choice that affects the success of a quail farming operation. Breeds differ in terms of size, egg production, and climatic adaptation. Among the several breeds available, the Coturnix quail is a popular option for commercial quail farming.

Coturnix quail, commonly known as Japanese quail, is recognized for their quick development, early maturity, and prolific egg production. These little birds come in a variety of colors, including white, brown, and speckled, giving farmers alternatives to meet their needs. Furthermore, Coturnix quails are noted for their kind demeanor, which makes them simpler to handle and maintain than certain other quail varieties.

When selecting a quail breed, consider climate, available area, and intended use. Bobwhite quails, for example, are popular among hunters because of their game bird traits, but Pharaoh quails are known for their adaptability to a variety of habitats.

Creating The Perfect Quail Habitat: Housing And Environment

To guarantee the success of a quail farming enterprise, it is essential to provide the birds with an appropriate environment. This entails providing enough housing and maintaining an environment that fosters the health and well-being of the quail population.

Quail housing should be built to keep the birds safe from harsh weather and any predators. While quail are resilient birds, providing shelter is critical for their safety and production. The design of quail homes varies, but important concerns include good ventilation, enough area, and ease of cleaning.

Ventilation is crucial to preventing the development of ammonia and moisture, which may cause respiratory problems in quail. Adequate space allows the birds to roam freely and show natural habits.

Furthermore, quick cleaning is vital for maintaining a sanitary atmosphere and reducing disease transmission.

The quail environment should closely resemble their native habitat as feasible. This involves supplying appropriate bedding, like straw or wood shavings, for nesting and dust bathing. Access to a well-balanced quail diet is critical to meeting their nutritional requirements for development, egg production, and general health.

Quails benefit from having a defined foraging area in their outdoor environment. This not only enables them to behave naturally, but it also helps to provide a more diverse and healthy food. Fencing is required to protect the quail from predators while still allowing them access to sunshine and fresh air.

To summarize, quail farming is an exciting option for farmers looking for a diversified and successful endeavor in the poultry market. Understanding the fundamentals of quail farming, choosing the appropriate quail breed, and developing an optimal environment are all critical aspects in assuring the success and sustainability of a quail farming enterprise.

CHAPTER TWO
Feeding Your Quails:
Nutritional Needs and Diet
Plan

Quails, tiny but prolific birds, need a well-balanced diet to live and produce nutritious eggs. Understanding their food needs is critical to keeping a healthy quail flock.

Nutritional Requirements

Quails need a diet high in protein, vitamins, and minerals. Their meals should include high-quality quail feed supplemented with grains and vegetables. Adequate protein is essential, particularly for laying hens, since it promotes egg production. Typically, a quail diet should have a protein concentration of 20-25%.

In addition to protein, quails need calcium to form robust eggshells. Oyster shells or crushed eggshells may be used as supplements. Vitamins and minerals, such as vitamins A, D3, E, and calcium, promote general health and bone formation. These critical elements are often included in a well-rounded commercial quail meal mix.

It's crucial to remember that water is an essential part of a quail's diet. Maintain a steady supply of clean, fresh water to keep your quails hydrated, particularly during hot weather or while they are producing eggs.

Diet Plans

Understanding your quails' particular demands at various life stages is essential when developing a balanced meal plan for them. Starter diets with a greater protein level (about 28-30%) are appropriate for chicks, giving the nutrients required for development. As they age, it is advised that they switch to a meal with a reduced protein concentration (20-25%).

Quails benefit from including greens and grains in their diet. This may contain spinach, kale, and grains such as millet and barley. These supplements not only diversify their nutrition but also improve their general health.

Quail Health Management: Prevention And Treatment Of Common Diseases

Maintaining a healthy quail flock requires aggressive health management to avoid common infections and handle difficulties quickly.

Preventive Measures

Quails are prone to a variety of illnesses, including respiratory infections and parasite infestations. Proper sanitation and cleanliness in quail housing are crucial for illness prevention. Regular cleaning of the coop, new bedding, and enough ventilation may dramatically lower the risk of illness.

New birds should be quarantined before being introduced to the current flock. This cautious approach helps to avoid the spread of dangerous illnesses. vaccine regimens are also available for some illnesses, and speaking with a veterinarian may help you develop a suitable vaccine plan.

Common Diseases And Treatment

Despite protective efforts, quails may still have health problems. Respiratory diseases, coccidiosis, and egg-laying issues are all prevalent concerns. Early identification is critical to successful therapy.

Common symptoms of respiratory infections include sneezing, coughing, and nasal discharge. Quarantine the infected birds and contact a veterinarian about suitable medications. Coccidiosis, a parasite illness, may be avoided with adequate cleanliness; nevertheless, if identified, treatments such as amprolium can be used.

Egg-laying issues, such as egg binding, need quick intervention. Providing calcium supplements and warm showers may help to alleviate the issue. Regular health inspections and timely action help to maintain a healthy quail flock.

Breeding Quails: Tips For Successful Reproduction

Successful quail reproduction requires careful planning and management to guarantee ideal breeding circumstances.

CHAPTER THREE

Selecting Breeding Stock

Selecting the appropriate breeding stock is critical to successful reproduction. Choose birds with ideal characteristics such as excellent health, high egg output, and appropriate temperament. To keep the quail population vigorous and healthy, avoid breeding birds with genetic defects or illnesses.

Creating Ideal Breeding Conditions.

Quails are prolific breeders when given the right circumstances. Maintain a constant light-dark cycle since quails are sensitive to day duration, which might induce reproductive activities. Adequate nesting boxes with soft bedding are required for pleasant egg laying.

It is critical to offer a well-balanced food throughout the breeding season, with an emphasis on increased protein intake to aid in egg production. Close monitoring of the breeding pairs ensures that any concerns are addressed as soon as possible.

Managing Quail Eggs: Collection, Incubation, And Hatching

A successful quail breeding enterprise relies on effective egg management from collection to hatching.

Egg Collection And Handling

To guarantee the freshness of quail eggs, they must be collected soon. Handle eggs with care, since harsh handling might result in cracks or damage. Before incubation, store eggs in a cool, dry place and identify them with the date to ensure freshness.

Incubation Process

Using a dependable incubator is essential for a successful hatching process. Maintain the ideal temperature and humidity conditions during the incubation phase. The eggs must be turned frequently to ensure consistent growth.

Candle the eggs regularly to assess embryo growth and remove those that are not viable. As the hatch date approaches, provide a clean and warm environment for the hatching process.

Hatching And Chick Care.

Once the eggs hatch, offer a safe and warm environment for the chicks. Pay attention to their first diet and a high-protein beginning meal. Adequate warmth, often given by heat lamps, is critical to the chicks' early growth.

Regular health checks and keeping an eye on the chicks' behavior improve their general well-being. As they age, gradually change their food and living arrangements to suit their growth.

Finally, knowing and adopting correct care, feeding, and management methods is critical for the effective maintenance of a quail flock. Quail aficionados may have a robust and prolific quail enterprise by emphasizing nutritional requirements, avoiding and resolving common health concerns, maintaining ideal breeding circumstances, and successfully managing the egg cycle.

CHAPTER FOUR

Raising Quail Chicks: Care And Development

Quail farming has grown in popularity as a sustainable and successful industry, thanks to the rising demand for quail goods.

Understanding the complexities of rearing quail chicks is critical for anyone interested in starting a quail farming operation. Successful quail chick development requires careful attention to their care and growth.

When it comes to rearing quail chicks, the first step is to provide a favorable atmosphere for their growth. The brooding place should be warm, dry, and safe. Using a brooder with a temperature control system is critical because quail chicks need a precise temperature range for optimum development. Adequate room, appropriate ventilation, and a consistent supply of clean water are all essential components of the quail chick habitat.

The growth of quail chicks is heavily dependent on feeding. A well-balanced quail starting feed high in protein is required for their first development. The meal should be finely powdered to accommodate the tiny size of the quail chicks. Furthermore, giving grit aids digestion and ensures that the chicks get enough nutrients from their meal.

Monitoring the health of quail chicks is critical in their early stages. Regular health checkups, immunization regimens, and timely response in the event of any indications of sickness are all necessary measures. Proper cleanliness and hygiene in the brooding area substantially help to avoid illness, resulting in a healthy and vigorous quail flock.

As quail chicks age, they must be moved to a grower feed. This feed should continue to provide their nutritional needs while supporting consistent growth and development. Paying attention to factors such as lighting and space availability is critical during this stage to avoid stress and promote a smooth transition to adulthood.

Quail Meat Production: Slaughter And Processing

The transition from quail chicks to meat production requires meticulous planning and execution. Slaughtering and processing are key parts of quail farming that have a direct bearing on the quality of the finished product.

Understanding the proper slaughter period is critical for producing high-quality meat. Quails are normally suitable for processing around 6 to 8 weeks old, achieving a mix of softness and size. Humane and efficient killing processes are required to reduce stress and maintain the well-being of the birds.

Once the quails have been killed, correct processing is essential to producing a high-quality product. To maintain sanitary requirements, feather removal, evisceration, and cleaning must be performed with care. Modern processing technology may help to simplify these operations, resulting in greater efficiency and uniformity in the end output.

The last stages of quail meat processing include packaging and storage. Proper packaging preserves the freshness of the meat and increases its shelf life. Whether selling directly to customers or via distributors, following packaging and labeling rules is critical for marketing and consumer confidence.

Marketing Quail Products: Strategies For Profitable Sales

In a competitive industry, good marketing methods are critical to the success of any quail farming operation. Creating a strong brand presence and targeting the proper audience are critical components of marketing your items for lucrative sales.

Identifying the unique selling qualities of your quail goods is the first step in developing an effective marketing campaign. Highlighting features such as meat quality, ethical agricultural procedures, or unique nutritional advantages may help your items stand out in the market.

Using internet channels and social media may dramatically increase the visibility of your quail items. Creating an appealing website, using social media networks, and even exploring e-commerce platforms may all help you engage directly with customers. Incorporating visually engaging information and telling the tale of your quail farm offers a personal touch that appeals to customers.

Collaboration with local markets, grocery shops, and restaurants opens up new opportunities for marketing quail goods. Building partnerships with chefs and culinary influencers may generate awareness for your business and attract a larger client base. Participating in farmers' markets and food festivals provides an opportunity for direct customer engagement and feedback.

Implementing promotions, discounts, and loyalty programs may help to increase repeat business and attract new clients. Word of mouth is a significant strategy, so guaranteeing client happiness and promoting reviews and testimonials may help you build a stronger market reputation.

CHAPTER FIVE
Record-Keeping And Financial Management In Quail Farming

A successful quail farming company involves not just a dedication to the birds, but also meticulous record keeping and financial administration. Keeping accurate records is critical for evaluating your quail flock's performance and making educated future choices.

Creating a complete record-keeping system entails maintaining several elements of your quail farm, including breeding data, health records, feed consumption, and production indicators. Regularly updating these records offers useful information about the general health and production of your quail farm.

In quail farming, financial management includes budgeting, cost analysis, and revenue projection. Creating a precise budget that includes expenditures like feed, veterinary care, infrastructure upkeep, and marketing initiatives aids in successful operational cost management. A cost-benefit analysis for several elements of your quail farming business will help you make better decisions and use resources more efficiently.

Monitoring income sources and seeking diversification options help your quail farming company remain financially sustainable. Understanding market trends, altering production levels depending on demand, and researching prospective value-added goods are all techniques that may help increase profitability.

In conclusion, successful quail farming requires a comprehensive strategy that includes quail chick care and growth, efficient meat production procedures, effective marketing tactics, and meticulous record-keeping and financial management.

By combining these aspects, quail farmers may not only fulfill market needs but also secure the long-term viability and profitability of their operations.

Troubleshooting Common Quail Farming Concerns

Quail farming has grown in popularity as a profitable industry owing to rising demand for quail eggs and meat. However, like with every agricultural venture, quail farming has its own set of obstacles.

Successful quail farming requires a thorough awareness of typical problems and the ability to troubleshoot successfully.

Inadequate Food And Nutrition

Quail farmers confront several issues, one of which is maintaining correct feeding and nutrition. Inadequate food may cause stunted development, decreased egg output, and general poor health in the quails.

To counter this, producers should give well-balanced and nutritious food. Commercial quail meals are available, however, adding green leafy vegetables and grains may improve their nutritional value.

Health Promotion And Disease Prevention

Quails are vulnerable to a variety of illnesses, and keeping them healthy is critical for a successful quail farming enterprise. Common illnesses include respiratory infections, coccidiosis, and dietary deficits. Regular health examinations and immunization programs are crucial. Immediately isolating unwell birds may help to avoid disease transmission. Proper sanitation and cleanliness in quail housing are critical components of illness prevention.

Insufficient Housing And Ventilation

Quails need proper habitat and ventilation to thrive. Inadequate housing can cause stress, lower egg production, and higher mortality. Quail houses should have adequate space, proper flooring, and good ventilation. Overcrowding should be avoided as it can cause aggressive behavior in quails. Adequate ventilation regulates temperature and humidity, lowering the risk of respiratory problems.

Scaling Up: Expanding Your Quail Farm to Increase Production

As a quail farming business grows, scaling up becomes an important strategic consideration. Expanding a quail farm requires careful planning and execution to ensure increased production without sacrificing product quality or bird well-being.

Infrastructure Development

Investing in infrastructure is required when expanding a quail farm. This includes building more housing units, providing enough nesting and laying areas, and making room for feeding and watering. Proper infrastructure development is critical to maintaining a favorable environment for quails to thrive.

CHAPTER SIX
Breeding Programs And
Genetic Improvement

Implementing a structured breeding program is critical for increasing production capacity. Selective breeding can improve desirable characteristics like egg-laying capacity, disease resistance, and growth rate.

Genetic improvement improves the overall efficiency and productivity of a quail farm. However, it's essential to strike a balance to avoid potential negative consequences, such as inbreeding depression.

Market Research And Product Diversification

Expanding a quail farm necessitates a thorough understanding of market dynamics. Conducting market research helps identify potential avenues for selling quail eggs and meat.

Diversifying products, such as offering processed quail products or value-added items, can open up new markets and revenue streams. Building strong relationships with local markets, grocery stores, and restaurants is crucial for sustained growth.

Sustainable Practices In Quail Farming

As the global focus on sustainable agriculture intensifies, quail farmers must explore and implement sustainable practices in their operations. Sustainable quail farming not only benefits the environment but also enhances the long-term viability of the business.

Environmentally Friendly Housing And Practices

Implementing environmentally friendly housing and management strategies is a cornerstone of sustainable quail farming. Utilizing renewable energy sources, such as solar electricity, may lower the carbon footprint of the farm. Proper waste management, including composting quail litter, promotes soil health and reduces environmental impact. Additionally, implementing water conservation techniques assures proper use of this crucial resource.

Natural And Organic Feeding Practices

Consumers increasingly choose goods sourced from animals grown on natural and organic diets. Sustainable quail farming entails adopting natural and organic feeding techniques. This may involve obtaining locally produced, non-

GMO feed, and avoiding the use of antibiotics or synthetic chemicals. Communicating these methods to customers may boost the commercial attractiveness of quail goods.

Management Of Biodiversity

Sustainable quail farming also requires measures to protect biodiversity. This involves protecting indigenous quail breeds and avoiding actions that lead to the decrease of wild quail populations.

Creating knowledge about the necessity of biodiversity protection among quail farmers develops a feeling of responsibility toward the greater ecological balance.

Conclusion

Quail farming is a feasible possibility for agricultural businesses seeking a specialized market. However, success in quail farming requires a mix of problem-solving abilities to meet frequent issues, strategic planning for expansion, and a dedication to sustainable techniques.

Farmers may provide the groundwork for a successful quail farming operation by addressing feeding, health management, and housing concerns. To address the increasing demand for quail goods, scaling up requires rigorous infrastructure construction, breeding programs, and market analysis. Adopting sustainable techniques protects the farm's long-term survival while also aligning with the worldwide trend toward environmentally aware agriculture. Finally, a comprehensive strategy that includes troubleshooting, scaling up, and sustainability is critical to thrive in the dynamic and gratifying world of quail farming.